Regenerative Farming: Nurturing the Land and Communities for a Sustainable Future.

The world is facing an unprecedented challenge of feeding a growing population while also preserving natural resources for future generations. Conventional agriculture, with its heavy reliance on synthetic inputs and industrial-scale practices, has led to soil degradation, water pollution, loss of biodiversity, and a myriad of other environmental and social problems. It is time to rethink the way we produce our food, and regenerative farming offers a promising alternative.

Regenerative farming is a holistic approach to agriculture that seeks to harness the power of nature to regenerate the health and productivity of the land. It is a system of farming that works with, rather than against, the natural ecosystem, utilizing natural resources in a way that they can regenerate their productive capacity and minimize harmful impacts on ecosystems beyond a field's edge. It prioritizes less resource-intensive farming solutions, greater diversity in crops and livestock, and farmers' ability to adapt to local circumstances.

In this book, we will explore the principles and practices of regenerative farming, and how they can help us build a more sustainable and resilient food system. We will delve into the latest research and case studies from around the world to illustrate the benefits of regenerative farming for the environment, public health, and local communities. We will also discuss the challenges and opportunities of transitioning from conventional to regenerative farming, and the role of farmers, policymakers, and consumers in promoting this transformation.

Ultimately, this book aims to inspire and empower readers to join the regenerative farming movement, whether as farmers, activists, or informed consumers. By working together to nurture the land and communities, we can create a more sustainable and equitable future for all.

I. Introduction

Definition Of Regenerative Farming

Regenerative farming is a holistic approach to agriculture that focuses on improving the health of the soil, increasing biodiversity, and enhancing ecosystem services. The goal of regenerative farming is not just to sustain productivity, but to actually regenerate the land, making it more fertile, resilient, and productive over time. This involves utilizing natural resources in such a way that they can regenerate their productive capacity, and also minimizing harmful impacts on ecosystems beyond a field's edge. Regenerative farming emphasizes less resource-intensive farming solutions, greater diversity in crops and livestock, and farmers' ability to adapt to local circumstances. This book explores the principles and practices of regenerative farming, and the ways in which it can help build a more sustainable future for both agriculture and society as a whole.

Importance Of Regenerative Farming For Sustainable Agriculture

Regenerative farming is increasingly gaining recognition as a critical component of sustainable agriculture. Regenerative farming seeks to promote soil health, enhance biodiversity, and reduce carbon emissions, making it a more ecologically responsible method of farming. By utilizing natural resources in a way that regenerates their productive capacity while minimizing harmful impacts on ecosystems beyond a field's edge, regenerative farming promotes a holistic approach to agriculture that benefits both the environment and communities. In this book, we will explore the principles and practices of regenerative farming, its importance for sustainable agriculture, and how it can help build a more resilient and sustainable future for our planet.

Brief History Of Regenerative Farming Practices

Regenerative farming practices are not new concepts. In fact, many traditional farming practices that were employed in the past can be considered regenerative in nature. For centuries, indigenous communities across the world have used sustainable farming methods that have helped preserve natural resources while providing food for their communities. Some of these practices include crop rotations, intercropping, and the use of natural fertilizers.

In the early 1900s, farmers in the United States began to adopt industrial farming practices that relied heavily on synthetic fertilizers and pesticides. While these practices increased crop yields, they also led to the degradation of soil health, water pollution, and other negative environmental impacts.

In recent years, there has been a renewed interest in regenerative farming practices as a way to address the negative effects of industrial agriculture. Today, many farmers and agricultural organizations around the world are turning to regenerative farming as a way to create a more sustainable food system that nurtures the land and communities.

II. Principles of Regenerative Farming

Utilizing Natural Resources For Productive Capacity

Regenerative farming seeks to utilize natural resources in a way that regenerates their productive capacity over time. By incorporating practices that build soil health, enhance biodiversity, and promote nutrient cycling, regenerative farmers work to create a resilient ecosystem that can support both healthy crops and thriving communities. This approach stands in contrast to traditional industrial agriculture, which often relies on synthetic inputs and degrades soil and other resources over time.

Regenerative farming focuses on improving soil health as the foundation of sustainable agriculture. Healthy soil is teeming with microbial life, which helps to break down organic matter and release nutrients to plant roots. By building up organic matter in the soil through practices like cover cropping, crop rotation, and reduced tillage, regenerative farmers can enhance soil health and improve yields over time.

In addition to improving soil health, regenerative farming practices also support biodiversity, reduce water usage, and sequester carbon in the soil. By minimizing the use of synthetic inputs and working with natural systems rather than against them, regenerative farmers can create a sustainable, regenerative ecosystem that supports healthy crops, healthy ecosystems, and healthy communities.

Minimizing Harmful Impacts On Ecosystems

Regenerative farming seeks to minimize harmful impacts on ecosystems beyond a field's edge, recognizing that farming practices can have significant effects on soil, water, air, and biodiversity. By adopting a holistic approach to land management, regenerative farmers strive to promote healthy soil, water quality, and habitat for wildlife while also producing crops and livestock. This approach recognizes that the health of the land, the plants, and the animals are all interconnected and interdependent. Through regenerative farming practices, farmers aim to create a self-sustaining system that works in harmony with nature, rather than against it.

Diversifying Crops And Livestock

Diversifying crops and livestock is a crucial aspect of regenerative farming. This involves growing a variety of crops and raising multiple types of animals on a single farm, which can help to improve soil health, increase biodiversity, and create a more resilient farm ecosystem. Monoculture farming, or the practice of growing only one crop, can lead to soil degradation, nutrient depletion, and increased susceptibility to pests and diseases.

In addition to crop diversity, regenerative farmers also utilize livestock as part of their farming systems. Animals can help to build healthy soils by grazing and fertilizing, and can also provide additional income streams for farmers through meat, dairy, and other animal products. By incorporating animals into their farming practices, regenerative farmers can create a more closed-loop system where waste products from one part of the farm become resources for another.

Overall, diversifying crops and livestock is a key strategy for regenerative farmers to improve the health and resilience of their farms, while also providing a range of benefits for the environment and local communities.

Adapting To Local Circumstances

Regenerative farming involves working with nature, rather than against it, to create resilient, productive and sustainable farming systems. It recognizes that the health of the soil, the quality of the water, and the diversity of plants and animals are all interconnected and essential to the long-term viability of the farm.

By utilizing natural resources in a way that regenerates their productive capacity, regenerative farming practices can help to restore degraded soils, mitigate climate change by sequestering carbon in the soil, increase biodiversity, and promote the health of ecosystems beyond the farm.

Regenerative farming also emphasizes the importance of diversity in both crops and livestock. By growing a variety of crops and integrating livestock into the farming system, farmers can reduce the risk of crop failure and improve soil health through natural fertilization.

Moreover, regenerative farming practices are adapted to local conditions and circumstances, including climate, soil type, and topography. This approach ensures that the farming system is well-suited to the local environment, reducing the need for external inputs such as fertilizers and pesticides.

This book will explore the principles and practices of regenerative farming, highlighting its importance for sustainable agriculture and offering practical guidance for farmers and land managers seeking to adopt these methods. Through case studies and examples from around the world, we will demonstrate the potential of regenerative farming to build healthy and resilient farming systems that benefit both farmers and the environment.

III. Regenerative Farming Practices

Agroforestry

Agroforestry is a land use management system that combines agricultural crops and/or livestock with trees or shrubs in a mutually beneficial way. Agroforestry practices can improve soil quality, increase biodiversity, conserve water, reduce erosion, and enhance resilience to climate change. This farming technique can also provide additional income streams through the sale of timber or non-timber forest products such as fruits, nuts, and medicinal plants. Agroforestry can be implemented in various forms, including alley cropping, silvopasture, and forest farming.

Cover Cropping

Cover cropping is a regenerative farming practice that involves planting non-cash crops during fallow periods or between cash crop cycles. Cover crops can improve soil health by adding organic matter, reducing soil erosion, suppressing weeds, and increasing soil fertility through nitrogen fixation. Additionally, cover crops can provide habitat for beneficial insects, birds, and other wildlife, which can help to reduce the need for pesticides. The choice of cover crop species depends on factors such as climate, soil type, and desired outcomes. Some common cover crop species include clover, rye, and buckwheat.

Crop Rotation

Crop rotation is a method of planting different crops in the same field over a series of growing seasons. This technique is used to maintain soil health and fertility, prevent soil erosion, control pests and diseases, and increase crop yields. By rotating crops, farmers can break pest and disease cycles, prevent the buildup of soil-borne pathogens, and reduce the need for synthetic fertilizers and pesticides. Additionally, different crops have different nutrient requirements, and rotating crops can help ensure that the soil remains balanced and healthy. Crop rotation has been practiced for centuries and is an important component of regenerative farming practices.

Regenerative Grazing

Regenerative grazing is a holistic approach to managing grazing animals, such as cows, sheep, and goats, in a way that mimics the natural grazing patterns of wild herds. This approach involves moving animals through a series of pastures, allowing them to graze on fresh vegetation and then move on to another area before overgrazing occurs. This allows for the regeneration of grasses and other vegetation, which can improve soil health, prevent erosion, and sequester carbon in the soil. Regenerative grazing also involves reducing or eliminating the use of synthetic fertilizers and pesticides, and incorporating other practices such as rotational grazing, cover cropping, and composting to build soil health and fertility.

Composting

Composting is a natural process that involves the decomposition of organic matter to produce a nutrient-rich soil amendment called compost. Composting can be done on a small scale, such as in a backyard compost pile, or on a larger scale, such as in a commercial composting facility. Composting helps to reduce the amount of organic waste sent to landfills, which in turn reduces the production of harmful greenhouse gases. Additionally, compost can improve soil health, increase water retention, and reduce the need for chemical fertilizers and pesticides, making it an important component of regenerative farming practices.

No-till farming

No-till farming is a method of growing crops without disturbing the soil through tillage, which is the mechanical agitation of the soil. Instead, the seeds are planted directly into the untilled soil, and the previous crop residues remain on the surface to act as a protective cover. No-till farming helps to build soil health by promoting the growth of beneficial microorganisms and reducing erosion. It also conserves water and energy, and reduces greenhouse gas emissions from the soil. No-till farming is often used in conjunction with other regenerative farming practices, such as cover cropping and crop rotation, to improve soil health and fertility.

Intercropping

Intercropping is a regenerative farming practice where two or more crops are grown in the same field at the same time, or in a sequence. Intercropping provides several benefits, such as reducing pests and diseases, enhancing soil health, and increasing yields. When plants with different growth habits are intercropped, they can also optimize the use of resources like sunlight, water, and nutrients. In addition to the ecological benefits, intercropping can also have economic advantages, as it can diversify income streams for farmers and reduce dependence on monoculture crops.

Polyculture

Polyculture is an agricultural system that involves growing multiple crops in the same field or plot of land at the same time, rather than monoculture, which is growing only one crop. In polyculture, different crops are chosen and combined in a way that they complement each other and create a more diverse and balanced ecosystem. This can result in increased soil fertility, reduced pest and disease pressure, and greater biodiversity. Polyculture can be practiced in both large-scale and small-scale agriculture, and can be used for both food and non-food crops. It is often used in regenerative farming practices to promote sustainability and resilience in agricultural systems.

IV. The Benefits of Regenerative Farming

Improved Soil Health And Fertility

Regenerative farming practices can greatly improve soil health and fertility, leading to healthier and more productive crops. By minimizing tillage and soil disturbance, soil structure is maintained, which helps to retain water, improve aeration, and increase the ability of soil to hold nutrients. The use of cover crops and crop rotations can also help to build soil organic matter, which is a key component of soil fertility.

Regenerative farming practices also focus on the use of natural fertilizers, such as compost and animal manure, rather than synthetic fertilizers. These natural fertilizers can improve soil health by increasing microbial activity, improving soil structure, and providing a more balanced mix of nutrients for plants.

Improved soil health and fertility can lead to greater yields and healthier crops, reducing the need for synthetic inputs and lowering overall costs for farmers. Additionally, healthy soils are better able to sequester carbon, reducing greenhouse gas emissions and contributing to the fight against climate change.

Increased Biodiversity

Regenerative farming practices have the potential to increase biodiversity on the farm by creating habitats for a variety of species. By diversifying crops and livestock and utilizing agroforestry and other land-use practices, regenerative farmers can create a more complex and interconnected ecosystem on their land. This can lead to an increase in beneficial insects, birds, and other wildlife, which can contribute to natural pest control and pollination, reducing the need for harmful pesticides and synthetic fertilizers. Additionally, by promoting a healthy soil ecosystem, regenerative farming practices can increase the populations of soil organisms, such as fungi, bacteria, and earthworms, which can improve nutrient cycling and soil structure. Overall, the increase in biodiversity associated with regenerative farming can contribute to a more resilient and sustainable agricultural system.

Carbon Sequestration

Carbon sequestration is the process of capturing and storing carbon dioxide from the atmosphere. In the context of regenerative farming, carbon sequestration can occur through various practices that promote healthy soil, such as cover cropping, crop rotation, and reduced tillage. When carbon is stored in the soil in the form of organic matter, it not only helps to mitigate climate change by reducing atmospheric carbon dioxide levels, but it also promotes healthy soil, which leads to improved crop yields, increased water retention, and increased biodiversity. In addition to promoting carbon sequestration through farming practices, regenerative agriculture also seeks to reduce carbon emissions by utilizing renewable energy sources and reducing fossil fuel use in farming operations.

Reduced greenhouse gas emissions

Regenerative farming practices have the potential to reduce greenhouse gas emissions from agricultural production. This can be achieved through several means, such as reducing tillage, which decreases soil disturbance and carbon loss from the soil. In addition, regenerative farming practices such as cover cropping and crop rotation can increase the amount of organic matter in the soil, which in turn can help sequester carbon. Regenerative grazing practices can also help reduce greenhouse gas emissions by improving soil health and increasing the amount of carbon stored in the soil. Overall, regenerative farming practices can contribute to mitigating the effects of climate change by reducing greenhouse gas emissions and increasing carbon sequestration in the soil.

Improved Water Quality And Conservation

Regenerative farming practices have been found to improve water quality and conservation. For example, cover crops and conservation tillage can reduce soil erosion and improve water infiltration, thereby reducing runoff and improving water quality. Regenerative grazing practices can also improve water quality by reducing erosion, compaction, and nutrient runoff from livestock manure.

In addition, regenerative farming practices can help conserve water resources. For example, intercropping and agroforestry systems can reduce water use by maximizing water use efficiency and reducing evaporation from the soil surface. No-till farming practices can also help conserve water by reducing soil disturbance and improving soil structure, which helps retain moisture in the soil. Furthermore, by increasing soil organic matter and improving soil structure, regenerative farming practices can help improve water-holding capacity and reduce the need for irrigation.

Economic Benefits For Farmers And Communities

Regenerative farming practices can have several economic benefits for farmers and communities, such as reducing input costs for fertilizers and pesticides, increasing yields, improving soil health and fertility, and diversifying income streams through the production of a range of crops and livestock. Additionally, regenerative farming can promote local food systems and strengthen community connections by fostering relationships between farmers and consumers. By prioritizing long-term sustainability and resilience, regenerative farming practices can also contribute to the economic stability and security of local communities.

V. Challenges and Limitations

Access To Resources And Information

Access to resources and information is essential for farmers who want to transition to regenerative farming practices. This includes access to funding, technical assistance, education, and training. Governments, non-governmental organizations, and private sector organizations can play a crucial role in providing these resources to farmers.

Funding can be provided through grants, loans, and subsidies to support farmers' investments in regenerative farming practices. Technical assistance can help farmers to implement these practices on their farms, troubleshoot issues, and learn about new developments in the field. Education and training can help to build farmers' knowledge and skills in regenerative agriculture, and support networks can provide farmers with opportunities to learn from one another and share best practices.

In addition to supporting individual farmers, access to resources and information can also benefit communities. Regenerative farming practices can improve soil health, enhance biodiversity, and reduce greenhouse gas emissions, leading to healthier ecosystems and cleaner air and water. They can also support local food systems and contribute to economic development in rural communities. By promoting regenerative farming practices and providing resources to farmers, we can create a more sustainable future for all.

Changing Consumer Demand And Supply Chains

Regenerative farming practices are not only important for the sustainability of the environment and the livelihoods of farmers, but also for the changing demands of consumers and supply chains. As consumers become increasingly conscious of the environmental impact of their purchases, there has been a growing demand for sustainably produced food. This demand has translated into an increased willingness to pay a premium for food that has been produced in a way that prioritizes regenerative farming practices.

At the same time, supply chains are recognizing the importance of sustainable and regenerative agriculture in meeting the demands of their customers. Companies that prioritize sustainable sourcing practices are more likely to attract and retain customers who value environmental sustainability, and are better positioned to navigate the regulatory and reputational risks associated with unsustainable practices.

Regenerative farming practices offer a means of meeting the growing demand for sustainable and ethically produced food, while also providing farmers with a means of improving their economic resilience and reducing their environmental impact.

Government Policies And Support

Government policies and support are critical for the widespread adoption of regenerative farming practices. Governments can provide financial and technical assistance to farmers to transition to regenerative farming methods. They can also create policies and regulations that support regenerative farming practices and promote the use of sustainable agricultural practices more broadly. Additionally, governments can provide education and training programs for farmers and the general public to increase awareness and understanding of regenerative agriculture and its benefits. Finally, governments can use their purchasing power to support regenerative agriculture by procuring food from regenerative farms for public institutions such as schools and hospitals.

Cultural Barriers And Traditions

Regenerative farming practices may face cultural barriers and traditional practices that have been passed down through generations. Some farmers may be hesitant to adopt new practices that differ from their traditions or ways of doing things. It is important to recognize the cultural importance of farming practices and to work with farmers to find ways to incorporate regenerative practices while respecting their traditions. Education and outreach programs that focus on the benefits of regenerative farming and its compatibility with traditional farming practices can help bridge this gap. Additionally, policies and financial incentives that support regenerative farming can encourage farmers to adopt these practices.

VI. Regenerative Farming Case Studies

Examples Of Successful Regenerative Farming Practices And Their Impacts

Certainly! Here are some examples of successful regenerative farming practices and their impacts:

Gabe Brown's Ranch: Gabe Brown, a farmer from North Dakota, has been practicing regenerative agriculture since the 1990s. By using practices such as cover cropping, no-till farming, and diverse crop rotations, Brown has been able to increase the organic matter in his soil, reduce erosion, and increase water infiltration. He has also been able to reduce his inputs, such as fertilizers and pesticides, and has seen an increase in biodiversity on his farm.

The Savory Institute: The Savory Institute is a global organization that promotes the practice of regenerative grazing. By using holistic management techniques, farmers are able to mimic the natural movement of large herds of grazing animals, which helps to improve soil health, water retention, and plant growth. This approach has been successful in restoring degraded landscapes in Africa, Australia, and the United States.

Polyface Farm: Polyface Farm is a family-owned farm in Virginia that uses regenerative farming practices to produce meat, eggs, and vegetables. By rotating livestock through different pastures and using a variety of grazing techniques, Polyface Farm has been able to improve soil health and increase the biodiversity of their land. They have also been able to reduce their reliance on fossil fuels and synthetic inputs, and have created a sustainable model for small-scale farming.

Singing Frogs Farm: Singing Frogs Farm, located in California, uses a combination of no-till farming, cover cropping, and composting to produce vegetables on a small scale. By focusing on building soil health, Singing Frogs Farm has been able to increase crop yields and reduce water use. They have also been able to reduce their carbon footprint and produce nutrient-dense food for their community.

Dorn Cox's Farm: Dorn Cox is a farmer in New Hampshire who has been practicing regenerative agriculture since the early 2000s. By using practices such as cover cropping, intercropping, and no-till farming, Cox has been able to increase soil health and biodiversity on his farm. He has also been able to reduce his inputs and has seen an increase in crop yields. Additionally, Cox has been working to develop open-source technologies to help other farmers adopt regenerative practices.

These examples demonstrate that regenerative farming practices can have significant positive impacts on soil health, water conservation, biodiversity, and economic sustainability. They also highlight the importance of local adaptation and experimentation in developing successful regenerative farming systems.

VII. The Future of Regenerative Farming

The Role Of Regenerative Farming In Addressing Global Challenges Such As Climate Change, Food Security, And Social Equity

Regenerative farming has the potential to address several global challenges. The agricultural sector is responsible for a significant portion of global greenhouse gas emissions, primarily through livestock production and the use of fossil fuels in farm operations. By implementing regenerative practices, farmers can reduce emissions by improving soil health, increasing biodiversity, and reducing the need for synthetic fertilizers and pesticides.

Regenerative farming can also contribute to food security by increasing the resilience of agricultural systems. Diversifying crops and livestock, improving soil health, and conserving water resources can improve the productivity and stability of farms, reducing the risk of crop failures and providing a more reliable food supply.

Furthermore, regenerative farming can promote social equity by creating opportunities for small-scale and marginalized farmers. By reducing the reliance on expensive inputs and improving yields, regenerative practices can provide a pathway for small-scale farmers to increase their income and improve their livelihoods.

Overall, regenerative farming offers a holistic approach to agriculture that addresses environmental, economic, and social challenges. As the world faces increasing pressures from climate change, population growth, and resource constraints, regenerative farming offers a promising solution for building sustainable and resilient food systems.

Emerging Technologies And Innovations In Regenerative Farming

Regenerative farming is a rapidly evolving field, with new technologies and innovations emerging all the time. Some of the most promising developments include:

Precision agriculture: This technology uses data and analytics to optimize farming practices, allowing farmers to reduce waste and increase yields while minimizing their environmental impact.

Soil sensors: By placing sensors in the soil, farmers can monitor factors like moisture, temperature, and nutrient levels in real time, allowing them to make more informed decisions about crop management.

Agroforestry: Combining trees and shrubs with crops and livestock can create a more diverse and resilient farm ecosystem, while also sequestering carbon and improving soil health.

Biodynamic farming: This holistic approach to farming involves incorporating spiritual and philosophical principles into agricultural practices, with a focus on regenerating the soil and creating a self-sustaining farm ecosystem.

Biochar: This carbon-rich material is produced by burning organic matter in the absence of oxygen, and can be used as a soil amendment to improve fertility and sequester carbon.

Cover crop cocktails: Planting multiple cover crops together can create a more diverse and resilient soil ecosystem, while also reducing erosion and improving soil health.

Regenerative grazing: Grazing livestock in a way that mimics natural patterns can improve soil health and biodiversity, while also reducing greenhouse gas emissions.

Aquaponics: This integrated system of fish farming and hydroponic agriculture can create a closed-loop ecosystem that requires minimal inputs and produces both protein and vegetables.

These and other emerging technologies and innovations have the potential to revolutionize agriculture, helping to create a more sustainable and regenerative food system for future generations.

Opportunities For Collaboration And Partnership

Collaboration and partnership are critical for advancing regenerative farming practices. Farmers, researchers, policymakers, and community members must work together to create a more sustainable and equitable food system. Some opportunities for collaboration and partnership include:

Farmer networks: Farmer networks provide a platform for farmers to share knowledge, resources, and best practices. These networks can be regional or national and can focus on specific crops, livestock, or regenerative practices.

Research partnerships: Researchers can work with farmers to study the impacts of regenerative practices on soil health, crop yields, and ecosystem services. These partnerships can help generate new knowledge and provide evidence-based support for regenerative farming.

Public-private partnerships: Public-private partnerships can bring together government agencies, non-governmental organizations, and private companies to support regenerative farming practices. These partnerships can provide funding, technical assistance, and other resources to farmers and communities.

Community-based organizations: Community-based organizations can work with farmers to promote regenerative practices and create local markets for sustainably produced food. These organizations can also help to increase access to healthy food in low-income communities.

Policy advocacy: Advocacy organizations can work to promote policies that support regenerative farming, such as conservation programs, organic agriculture standards, and local food procurement policies. By advocating for policy change, these organizations can help to create a more supportive environment for regenerative farming practices.

Overall, collaboration and partnership are essential for promoting regenerative farming practices and building a more sustainable and equitable food system.

VIII. Conclusion

Recap Of Key Points

Regenerative farming is an approach to agriculture that seeks to utilize natural resources in such a way that they can regenerate their productive capacity and minimize harmful impacts on ecosystems beyond a field's edge. It involves less resource-intensive farming solutions, greater diversity in crops and livestock, and farmers' ability to adapt to local circumstances. Regenerative farming practices include agroforestry, cover cropping, crop rotation, regenerative grazing, composting, no-till farming, intercropping, and polyculture.

The benefits of regenerative farming practices include

improved soil health and fertility, increased biodiversity, carbon sequestration, reduced greenhouse gas emissions, improved water quality and conservation, and economic benefits for farmers and communities. However, there are also barriers to widespread adoption, including access to resources and information, changing consumer demand and supply chains, government policies and support, and cultural barriers and traditions.

Examples of successful regenerative farming practices demonstrate the positive impacts on soil health, biodiversity, and economic benefits for farmers and communities. Regenerative farming can also play a crucial role in addressing global challenges such as climate change, food security, and social equity. There are emerging technologies and innovations in regenerative farming, and opportunities for collaboration and partnership to further advance the field.

Call To Action For Individuals, Policymakers, And The Farming Community To Support Regenerative Farming Practices

Regenerative farming has the potential to transform our agricultural systems, address global challenges such as climate change and food insecurity, and foster more sustainable and equitable communities. However, achieving this vision requires a collective effort from individuals, policymakers, and the farming community.

Individuals can support regenerative farming by choosing to purchase products from farmers who use regenerative practices, supporting local food systems, and advocating for policies that promote sustainable agriculture.

Policymakers can play a critical role in promoting regenerative farming practices by investing in research and development, providing financial incentives for farmers, and creating policies that support sustainable agriculture.

The farming community can lead the way in adopting regenerative practices, sharing knowledge and experiences, and collaborating with others to promote a more sustainable and equitable food system.

By working together, we can build a more sustainable and regenerative future for our communities, our land, and our planet. Let us take action now to support regenerative farming practices and create a more just and sustainable world for future generations.

A s we have explored throughout this book, regenerative farming offers a holistic approach to sustainable agriculture that nurtures the land and communities for a sustainable future. By utilizing natural resources in a way that regenerates their productive capacity and minimizing harmful impacts on ecosystems beyond a field's edge, regenerative farming presents a promising solution to the pressing global challenges of climate change, food security, and social equity.

Through diverse practices such as agroforestry, cover cropping, crop rotation, regenerative grazing, composting, no-till farming, intercropping, and polyculture, regenerative farming can improve soil health and fertility, increase biodiversity, sequester carbon, reduce greenhouse gas emissions, improve water quality and conservation, and generate economic benefits for farmers and communities.

To fully realize the potential of regenerative farming, collaboration and partnerships are necessary between individuals, policymakers, and the farming community. We must also overcome cultural barriers and traditions and leverage emerging technologies and innovations to scale up regenerative farming practices.

As we move forward, we must continue to raise awareness about the importance of regenerative farming and advocate for government policies and support. By embracing regenerative farming practices, we can work towards a more sustainable and equitable future for all.

www.ingramcontent.com/pod-product-compliance
Ingram Content Group UK Ltd.
Pitfield, Milton Keynes, MK11 3LW, UK
UKHW020630010625
6176UKWH00042B/3551

9 798393 500603